I0483966

Centralized vs Decentralized Air Conditioning Systems

Abstract

Commercial air conditioning may be provided by a variety of equipment ranging from low horsepower self-contained systems to the very large built-up central systems of several thousand ton capacity. Customer/user's ultimate objective is to acquire and utilize an air conditioning system that will provide the most appropriate performance on a whole of life basis, in terms of capital, operating, replacement and maintenance costs. It's the architect's/HVAC engineer's responsibility to guide and advise the customers the best option.

HVAC systems are of great importance to architectural design efforts for four main reasons.

1. First, the success or failure of thermal comfort efforts is usually directly related to the success or failure of a building's heating, ventilation and air conditioning (HVAC) systems;

2. Second, HVAC systems often require substantial floor space and/or building volume for equipment and distribution elements that must be accommodated during the design process;

3. Third, HVAC systems require significant capital investments;

4. Last, but not least, the HVAC system is responsible for large portion of building operating costs.

The design and selection of right HVAC system therefore must combine a proper choice of engineered products efficiently providing conditioned air to the space at optimum energy while adding architectural features that shall complement the interior design.

This 5 - hr Quick Book course discusses the various issues to be considered and the questions to be raised before an intelligent, well-thought HVAC scheme is finalized. The advantages and disadvantages, which arise as a result of centralized or decentralized air conditioning systems, are evaluated in this course.

The course is intended for use primarily to the architects, engineers, contractors, facility managers, O&M personnel and HVAC designers.

Learning Objective

At the conclusion of this course, the student will:

- Understand the different types of HVAC systems;

- Understand the HVAC design challenges on various architectural and functional parameters;

- Understand the difference between central and decentralized systems;

- Understand the central chilled water system sub-configurations;

- Understand various types of decentralized systems (window, split, package, heat pumps and VRF systems);

- Understand the pros and cons of the central and compact systems;

- Understand how to select the right system for your application.

PART 1   **TYPE OF AIR CONDITIONING SYSTEMS**

Investment in a building project entails significant capital investment and associated costs over the economic life of the project. It is a mistaken notion that the buildings costs have to be expensed once. The buildings like any other industry have running expenses in a way that they consume lot of energy and require water and disposal facilities that accounts for significant recurring costs.

While there are many different HVAC system designs and operational approaches to achieving proper system functionality, every building is unique in its design and operation. For instance residential apartments, shopping complex, office complex, hospital, hotel, airport or industry; all have different functional requirements, occupancy pattern and usage criteria. The geographical location of the building, ambient conditions, indoor requirements, building materials, dimensional parameters, aesthetic requirements, noise and environment issues need different treatment. The HVAC design options must therefore be customized to satisfy the owner's business needs, architect's vision combined with operational and maintenance requirements of the facilities manager.

The selection of most appropriate HVAC system is related to various parameters, including but not limited to:

Parameters	HVAC Challenges
Thermal Comfort	The internal environment of the buildings must be a major focus point in the HVAC system selection. A number of variables interact to determine whether people are comfortable with the temperature of the indoor air. The activity level, age, and physiology of each person affect the thermal comfort requirements of that individual. The American Society of Heating, Refrigerating, and Air-Conditioning Engineers (ASHRAE) Standard 55-1981 describes the temperature and humidity ranges that are comfortable for most people engaged in largely

Parameters	HVAC Challenges
	sedentary activities.
Building Architecture	The HVAC system selection is influenced by the characteristics of the building such as: • Purpose of the building, area classification, occupancy and usage patterns; • Type of building structure, orientation, geographical location, altitude, shape, modules- size and height; • Materials and thickness of walls, roof, ceilings, floors and partitions and their relative positions in the structure, thermal and vapor transmittance coefficients, areas and types of glazing, external building finishes and color as they affect solar radiation, shading devices at windows, overhangs, etc.; • Ratio of internal to external zones, glazing, plant room sitting, space for service distribution; • Foundation and supports requirement, permissible loadings.
Available Space	Considerable space is needed for mechanical rooms to house the HVAC equipment. In addition shaft spaces are required for routing ducts/pipes and other services e.g. electrical and plumbing work. Early liaison is therefore required with the

Parameters	HVAC Challenges
	project architect to proportion the building that would be occupied by HVAC systems, as this will have an impact on the size and cost of the building.
Building ceiling heights	The HVAC designer must thoroughly evaluate the ceiling space for air distribution ducts. Low ceiling height clearance between suspended ceilings and roof (also called plenum space) require close co-ordination with structural group for location and size of structural beams. Inadequate spaces to run ducts, probably force the system designer to use decentralized or unitary air conditioning units.
Building Aesthetics	The HVAC layout should be complementary to the building architecture. Often the requirements are stringent for example: • No equipment should be visible or should suitably blend with environment • Size and appearance of terminal devices in ceiling shall harmonize with lighting layout, fire sprinklers, detectors, communication systems and ceiling design; • Acceptability of components obtruding into the conditioned space; • Accessibility for installation of equipment, space for maintenance;

Parameters	HVAC Challenges
	• Location of fresh air intakes and exhausts, fire zones and fire walls; • Indoor and outdoor equipment preferences etc.
Efficiency/Performance and Energy Use	To assemble the best HVAC system, the efficiency, performance, cost and energy use will be major considerations when selecting components for the system. The equipment selected must conform or excel beyond the constraints presented in ASHRAE Standard 90.1. The cost of the energy consumed by the components of the HVAC system is an important aspect of the system selection. Each component must use as little energy as possible and still meet the performance requirements. Efforts should be on improving the building Green Star rating and obtaining Leadership in Energy and Environment Design (LEED) certification. Issues such as integration of the proposed system with the Building Management System, existing plant types etc must be taken into consideration as well.
Availability of water	The places where water is scarce, the only choice leans towards air-cooled equipment. In other public places such as hospitals, the air cooled equipment is sometime preferred due to potential concerns of Legionalla disease

Parameters	HVAC Challenges
	associated with water cooling.
Noise control	Sufficient attenuation is required to minimize equipment and air distribution noise. It is important to select low decibel equipment and define its location relative to the conditioned space.
Indoor environment and its control	Equipment and control design must respond to close tolerances on temperature/humidity, cleanliness, indoor air quality etc. Zone control or individual control is important consideration for the anticipated usage patterns. HVAC system design, particularly control system shall not be over-reliant on user interaction to operate the system. Other than maintaining zone or individual temperature and humidity conditions in permissible tolerances, the control and operational requirements include – supervision, records, type of adjustment and regulation, hours of operation, summer/winter changeover, day/night and weekend operation, high/low limit protection, frost protection, fire protection, special control areas (e.g. computer rooms, executive offices).
Environmental constraints	The refrigerant technology is changing because of concerns of ozone depletion. The production of several commonly used refrigerants will soon end, including R-22 in 2010 and R-123 in 2020 in accordance to the requirements of the Montreal Protocol. Every effort should be made to specify

Parameters	HVAC Challenges
	equipment which does not require any chlorofluorocarbons (CFC) refrigerants, including R-11, R-12, R113, or R114, or R-500. Hydrochlorofluorocarbons (HCFC) refrigerants such as R-22 and R123 are discouraged. By considering the phase-out of CFC refrigerants and fast approaching deadlines for HCFC refrigerants, the recommended refrigerants should be **HFC** such as R134a or azetropes R407c or R404a where possible. Cooling equipment that avoids CFCs and HCFCs eliminates a major cause of damage to the ozone layer.
Robustness & Redundancy	Consideration shall be given to the appropriate level of reliability of the HVAC plant, suitable redundancies, the consequences of failure of an item of plant, and alarm notifications particularly for facilities with high Occupational Health and Safety (OH&S) and technological risks/requirements (e.g. calibration centers).
Delivery and Installation schedules	HVAC designer must evaluate the equipment options that provide short delivery schedules and are relatively easy to install.
Type of ownership	Since a single building may have a single owner or multiple owners, energy billing, maintenance, usage timing is an important aspect in multiple ownership.

Parameters	HVAC Challenges
System flexibility	The HVAC designer need to consider the likelihood of space changes. It is likely that from time to time the users may need or wish to change the layout of rooms or the intended use of the internal environment. They may prefer systems that facilitate this to minimize consequent disruption and cost. Some systems lend themselves to adaptation of cellular layouts or from open plan to cellular.
Codes & Standards	The selection of the HVAC system is often constrained by various local codes and ASHRAE standards. As a minimum, the design shall follow: • ASHRAE Standard 15: Safety Standard for Refrigeration Systems, • ASHRAE Standard 55: Thermal Environmental Conditions for Human Occupancy, • ASHRAE Standard 62: Ventilation for Acceptable Indoor Air Quality, and • ASHRAE Standard 90.1: Energy Standard for Buildings except Low-Rise Residential Buildings. In addition the HVAC system selection is influence by statutory standards: • Government and local regulation on

Parameters	HVAC Challenges
	occupancy & safety classification;
	• Regulations of Public utilities on electrical wiring, power usage, water supply and drainage;
	• Health and Safety regulations on indoor air quality, ventilation air quantities, noise control, electrical, fuel, insulation and other hazardous materials;
	• Local fire authority regulations and smoke removal systems;
	• Insurance company regulations.
Life cycle costs	Capital, running costs, maintenance costs, and plant replacement costs need to be taken into account so that the selected system demonstrates value for money to install and operate. An important consideration is how much energy is used by a system and energy optimization measures need to be assessed during the life cycle costing process.

These and many other variables factor into the decision-making process. Bringing all of these constraints to a common solution requires sound knowledge of available technology, skillful evaluation of HVAC options, scrutinizing them and ultimately selecting the best alternatives. We will learn this more in part-II of the course.

Types of HVAC Systems

There are several choices for the type of air conditioning systems, each satisfying the HVAC objectives with different degrees of success. Broadly the air conditioning system can be classified in two broad categories: 1) Centralized air conditioning systems and 2) Decentralized systems.

- Central air conditioning systems serve multiple spaces from one base location. These typically use chilled water as a cooling medium and use extensive ductwork for air distribution.

- Decentralized air conditioning systems typically serve a single or small spaces from a location within or directly adjacent to the space. These are essentially direct expansion (DX)* type and include:

  o Packaged thru-the-wall and window air conditioners;

  o Interconnected room by room systems;

  o Residential and light commercial split systems;

  o Self-contained (floor by floor) systems;

  o Commercial outdoor packaged systems

*In DX refrigeration the air is cooled directly exchanging heat from the refrigerant.

The principal advantages of **central** air conditioning systems are better control of comfort conditions, higher energy efficiency and greater load-management potential. The main drawback is that these systems are more expensive to install and are usually more sophisticated to operate and maintain.

The principle advantages of **decentralized** air conditioning systems is lower initial costs, simplified installation, no ductwork or pipes, independent zone control, and less floor space requirements for mechanical room, ducts and pipes. A great benefit of decentralized systems is that they can be individually metered at the unit. Disadvantages are short equipment life (10 years), higher noise, higher energy consumption (kW/ton) and are not fit where precise environmental conditions need to be maintained.

CENTRAL SYSTEMS

Centralized systems are defined as those in which the cooling (chilled water) is generated in a chiller at one base location and distributed to air-handling units or fan-coil units located through out the building spaces. The air is cooled with secondary media (chilled water) and is transferred through air distribution ducts.

A typical chilled water central system is depicted in Figure below. The system is broken down into three major subsystems: the chilled water plant, the condenser water system (or heat rejection system) and the air-delivery system.

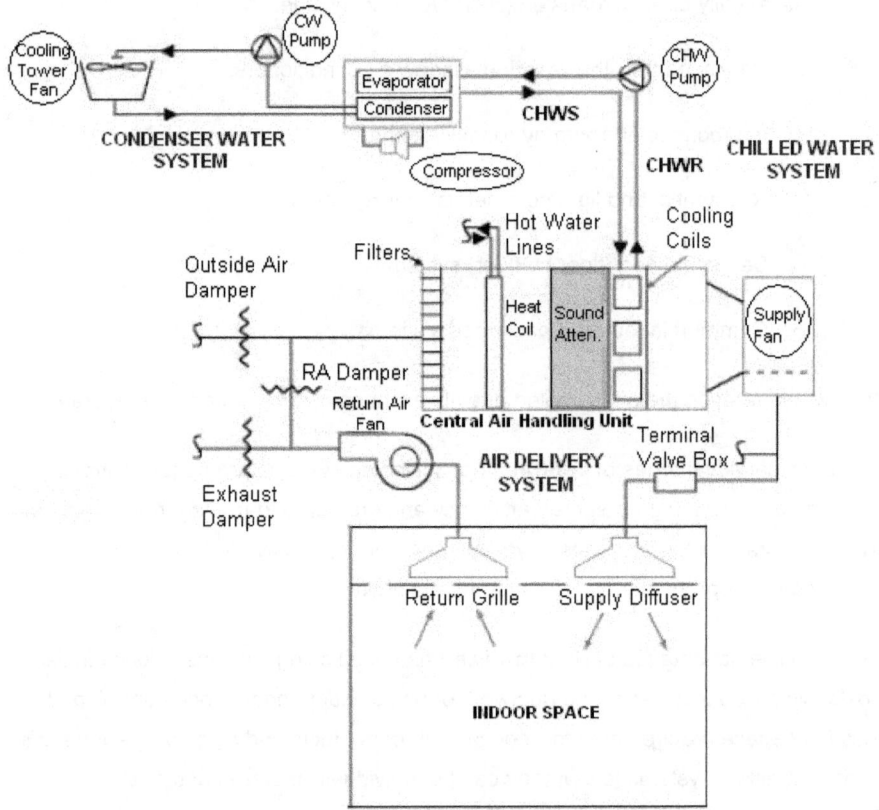

Chilled Water System: The chilled water system supplies chilled water for the cooling needs of all the building's air-handling units (AHUs). The system includes a chilled water pump which circulates the chilled water through the chiller's evaporator section and through the cooling coils of the AHUs. The system may have primary and secondary chilled water pumps in order to isolate the chiller(s) from the building: the primary pumps ensure constant chilled water flow through the chiller(s), while the secondary pumps deliver only as much chilled water is needed by the building AHUs.

Three most common chillers options are - reciprocating compressors (up to 200 TR*), screw compressors (100 to 750 TR) and centrifugal compressors (200 to 2000 TR). The centrifugal compressors offer the best peak load efficiency while screw chillers give better part load and the off-design performance.

[**TR*** stands for Ton of Refrigeration and is defined as the ability of the air-conditioning equipment to extract heat. 1TR is equal to heat extraction rate of 12000 Btu/h].

Condenser Water System: A refrigeration system must also reject the heat that it removes. There are two options for heat rejection: 1) air cooled and 2) water cooled.

- **Air cooled units** absorb heat from the indoor space and rejects it to ambient air. Air cooled units incorporate a condensing unit comprising of condenser, compressor, propeller fans and controls assembled in one unit and located outdoors. These are the most common system used in residential and light commercial applications.

- **Water cooled units** absorb the heat from the indoor space and rejects that heat to water which in turn may either reject heat via fluid coolers or cooling towers, or dry air coolers with adiabatic kits. Due to the lower refrigerant condensing temperatures compared to air cooled systems, water cooled chillers have higher coefficient of performance (COP). These are most common where good quality water is available and for large buildings such as multistory offices, hotels, airports and shopping complexes.

Air Delivery System: Air is drawn into a building's HVAC system through the air intake by the air handling unit (AHU). Once in the system, supply air is filtered to remove particulate matter (mold, allergens, and dust), heated or cooled, and then circulated throughout the building via the air distribution system, which is typically a system of supply ducts and registers.

In most buildings, the air distribution system also includes a return air system so that conditioned supply air is returned to the AHU ("return air") where it is mixed with supply air, re-filtered, re-conditioned, and re-circulated throughout the building. This is usually accomplished by drawing air from the occupied space and returning it to the AHU by: 1) ducted returns, wherein air is collected from each room or zone using return air devices in the ceiling or walls that are directly connected by ductwork to the air-handling unit; or 2) plenum returns, wherein air is collected from several rooms or zones through return air devices that empty into the negatively pressurized ceiling plenum (the space between the drop ceiling and the real ceiling); the air is then returned to the air-handling unit by ductwork or structural conduits. Finally, some portion of the air within is exhausted from the building. The air exhaust system might be directly connected to the AHU and/or may stand-alone.

System Types

The Central system category could be further broken down into the following:

- Central systems with CAV air-handling units
- Central systems with VAV air-handling units
- Central systems with fan-coil units (All- Water systems).

Constant air volume (CAV) system is an all-air system which accomplish cooling and heating by varying the supply air temperature and keeping the air volume constant. The system works well and maintains comfortable conditions in spaces with uniform heating and cooling requirements.

Variable Air Volume (VAV) system is an all air system which can satisfy the individual cooling requirements of multiple thermal zones. This is achieved by supplying air at a constant temperature from central plant to one or more VAV terminal units in each zone and adjusting the amount of supply air to meet required cooling loads.

The primary benefit of VAV over constant volume systems (CV) is its ability to simultaneously provide the required level of cooling to any number of zones within a building.

Key points

- Used in buildings with multiple zones to match the particular cooling/heating demands of each zone.

- Can be relatively energy efficient due to the ability to reduce the speed of the supply/extract fan(s) during periods of low to moderate loads.

Limitations

- Design and commissioning is particularly important if good system performance is to be achieved in terms of comfort and energy efficiency.

- Fan-assisted terminal units generally have higher capital and maintenance costs and the potential for increased noise levels.

- The designer needs to ensure adequate outside air is provided when the VAV terminal is regulated down to off set moderate thermal cooling loads.

- The designer needs to take care with the air distribution equipment to ensure dumping of supply air does not occur when the VAV terminal is regulated down to suit moderate cooling loads.

- Fan assisted VAV units do not adequately filter the recirculated air

All-Water Systems: Central all-water systems with fan-coil units use un-ducted arrangement. Here chilled water is pumped from the central plant through pipes to the fan coil terminal units placed inside the conditioned space. The room air is re-circulated through the unit and is cooled by the coil. Fan coils are available in a range of sizes, but can be broadly divided between the perimeter under-window console type and ducted units generally installed in a ceiling space.

Fan coils offer many benefits including good environmental control and air movement however have increased maintenance requirements compared with a "all-air" ducted system and require maintenance access to the occupied space. Each unit contains a filter which requires regular cleaning/changing. Generally, fan coils are quiet, but noise can be a problem in some situations where the fans are close to the conditioned space, and appropriate acoustic treatment needs to be considered.

Limitations

- Each fan coil unit incorporates a filter which requires regular cleaning/changing which can be difficult to access.

- There is a risk of water leaking from overhead fan coils into the space below.

- Floor mounted perimeter fan coils can occupy valuable floor space.

- Potential noise issues due to short duct runs from the supply air fan to the air conditioning outlets.

Typical Applications of Central Systems

Centralized systems are mostly used in mid to high rise buildings, which are structures with 5-7+ floors. Commercial buildings commonly choose several types of systems based on the space conditioning needs of different systems. A constant-volume (CV) system might cool the interior, which has relatively uniform cooling requirements while a VAV system conditions perimeter areas, which have variable requirements. Table below shows some typical applications for various types of systems.

Building Type	Type of System
Office Buildings (low rise)	VAV; or CV in the core, and hydronic at perimeter
Office Buildings (high-rise)	Central CV system for core and VAV or hydronic at perimeter
Department Stores	Multiple CV or VAV air handlers
Universities	CV, VAV or combined air-water systems at each building
Schools	CV or VAV air handlers serving individual common areas, and hydronic or combined air-water systems in classrooms
Hospitals	Separate CV systems for critical areas; CV or VAV for common areas; hydronic and combined air-water in patient rooms
Hotels	VAV for common areas like lobbies, restaurants, ball rooms & banquets; fan-coil units in guest rooms for individual temperature and humidity control
Assembly, Theatres	Multiple VAV air handlers
Libraries, Museums	Multiple CV air handlers, with precise humidity and temperature control

Central systems are also available as DX systems but in true sense these are large split systems. For example in a multistory building above 6 floors, chilled water system can work with chillers located at one central location (in basement or ground level) and the cooling is achieved by circulating chilled water through various air handling units located at multiple floors. For DX system there is limit to the length of refrigerant piping and the best configuration may be achieved by incorporating individual localize DX system for each floor. We will discuss this further in subsequent sections.

DECENTRALIZED SYSTEMS

Decentralized air conditioning systems commonly known as by various generic names viz. local systems, individual systems, floor-by-floor systems, unitary systems or packaged systems provide cooling to single room/spaces rather than the building. These are also referred to as "Direct Expansion" or DX types since the cooling is delivered by exchanging heat directly with a refrigerant type cooling coil and these do not use chilled water as an intermediate cooling medium.

These units are factory designed modular units all assembled into a package that includes fans, filters, heating source, cooling coil, refrigerant coils, refrigerant side controls and condenser. All cooling and heat rejection occur within the envelop of the unit. Each component is matched and assembled to provide specific performance specifications.

Window Air Conditioner

Window air conditioner provides cooling only when and where needed and is less expensive to operate. In this air conditioner all the components, namely the compressor, condenser, expansion valve or coil, evaporator and cooling coil are enclosed in a single box which is fitted in a slot in the wall of the room, or often a window sill. Room air conditioners are generally available in capacities varying from about 0.5 TR to 3 TR.

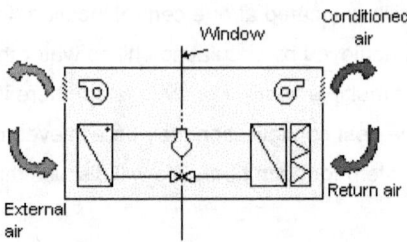

Typical Window Unit

Split Air conditioning Systems

The split air conditioner comprises of two parts: the outdoor unit and the indoor unit. The outdoor unit, fitted outside the room, houses components like the compressor, condenser and expansion valve. The indoor unit comprises the evaporator or cooling coil and the cooling fan. The indoor and outdoor units are connected by refrigerant pipe that transfers the refrigerant. Separation distance between exterior and interior elements is usually limited to around 100 feet. Split-systems are popular in small, single-story buildings. For this unit you don't have to make any slot in the wall of the room.

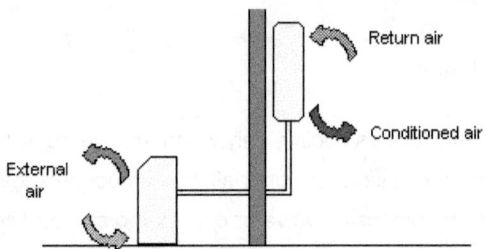

Typical Split Unit Arrangement

Flexibility is the overriding advantage of a split system. Because a split system is connected through a custom designed refrigerant piping system, the engineer has a large variety of possible solutions available to meet architectural and physical requirements particularly for buildings with indoor and/or outdoor space constraints. For example, the evaporator unit might be located in a basement; interior closet or attic while the compressor/ condenser unit might be located on the side, rear or roof of a building.

Variable Refrigerant Flow (VRF) Split System

A VRF air-conditioning system is essentially a sophisticated split system with an added ability to provide cooling on an individual basis to multiple rooms from a common condenser. Central to VRF control is their ability to automatically vary refrigerant flow in response to the heating/cooling load of the building. Occupant control is very simple, with easy to use wall-mounted key pads or hand held remote controllers providing individual control of room units. This is particularly useful in applications such as office blocks, hotels and large retail stores etc. which may need cooling in some areas and heating in other areas.

VRF systems are complex and contain microprocessor-based electronics, which ensure efficient operation and simple individualized control. Draw back is that these systems can have longer refrigerant piping runs and significant amount of refrigerant passes through occupied spaces. This could potentially cause a problem if a leak occurs.

Packaged Air Conditioners

Packaged HVAC systems consist of pre-assembled, off-the-shelf equipment that provides space heating, cooling, and ventilation to small and medium spaces. An HVAC designer will suggest package type of air conditioner if you want to cool more than two rooms or a larger space at your home or office. Packaged air conditioning systems are available in capacities ranging from about 5 TR to up to about 100 TR and a standard package unit is typically rated at 400 CFM (cubic feet per minute) supply air flow rate per ton of refrigeration. Obviously the larger the tonnage, the larger will be the airflow and it will require ductwork to cover all spaces and to reduce noise.

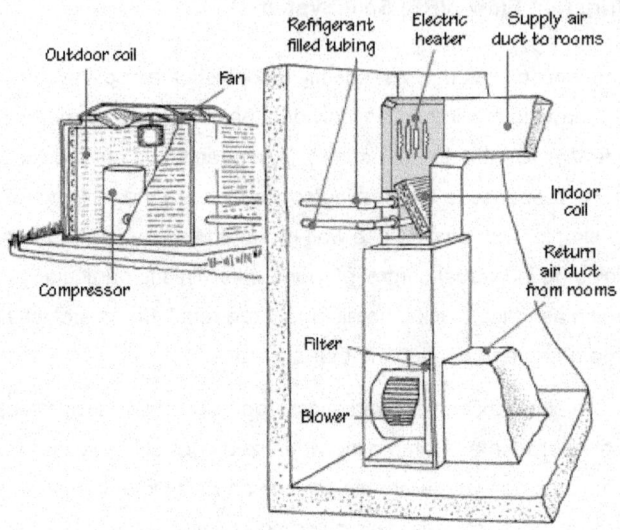

Package Type Split System

Package terminal air conditioners (PTAC):

Package terminal air conditioners (PTAC) also called "through-the-wall" air conditioners are relatively small systems typically below 7.5 TR and require no external ductwork. They are like a commercial quality version of residential window-mounted air conditioners (although they are actually mounted at floor level in a sleeve passing through the building wall).

Ductless products are fundamentally different from ducted systems in that heat is transferred to or from the space directly by circulating refrigerant to evaporators located near or within the conditioned space. In contrast, ducted systems transfer heat from the space to the refrigerant by circulating air in ducted systems.

Single package rooftop systems

These systems consist of a single rooftop-mounted unit that contains all mechanical elements of the HVAC system, including compressors, condensers, and evaporators. The units also include a supply fan and filter system that connects to the ductwork to provide air to the conditioned space or can be used with air distribution ductwork.

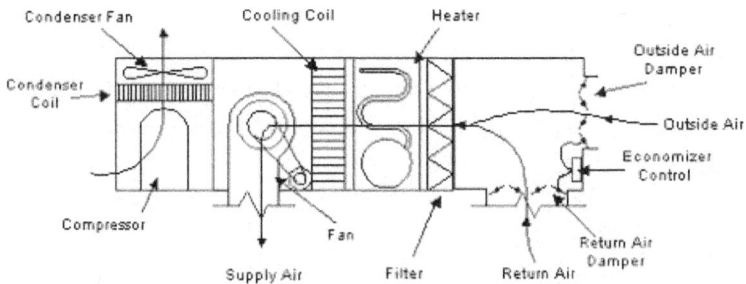

Typical Single-Package Rooftop System

The typical capacity for a rooftop-packaged unit is 5 to 130 tons. Rooftop units work well for single-story buildings, but don't fit into multi-storey schemes. These units are popular for general air-conditioning of stores, residences, schools, offices, etc. particularly suitable for single flat building with extensive floor areas

Heat Pumps

Heat pumps are similar to cooling only systems with one exception. A special 4-way valve in the refrigeration piping allows the refrigeration cycle to reverse so that heat is extracted from outside air and rejected into the building. Heat pumps provide both heating and cooling from the same unit and due to added heat of compression, the efficiency of heat pump in heating mode is higher compared to the cooling cycle.

In the summer heat pumps work like a standard air conditioner removing heat from inside your home and transferring it to the outside through the condenser coil. In the winter heat pumps run in reverse removing heat from the outdoor air and transferring into the home by the evaporator coil, which now becomes a condenser coil in the heating mode. As the temperature drops outside, the unit must work harder to remove heat from the air, lowering its efficiency. At this point, a heat pump system will use supplemental electric resistive heaters to warm the air to the proper temperature, similar to the heating elements in a toaster.

Heat Rejection

Most decentralized systems use air-cooled finned tube condensers to expel heat. The larger packaged air conditioners may be water cooled or air cooled.

Typical Applications

Decentralized systems are used in most classes of buildings, particularly where low initial cost and simplified installation are important, and performance requirements are less demanding.

Building Type	Type of System
Residences, Dormitories	Window or Split Units, Heat Pumps or Package Units
Office Buildings (low rise)	Split Units, Package Units, Roof top Units
Department Stores	Rooftop Units, Package Units
Restaurants	Package Units
Motels	Package Units, Split Units, Heat Pumps, Roof top Units
Small commercial complexes	Package Units, Rooftop Units
Cinema Halls, Theatre	Rooftop Units, Package Units, Custom built DX Units
Library	Rooftop Units, Package Units, Custom built DX Units
Medical centers, clinics	Rooftop Units, Package Units, Custom built DX Units

Note on Roof top and Package Units:

Decentralized systems are considered as standard off shelf catalogue products, which include large split system, the roof top units and the cabinet package units. Despite not being distant from the rooms they have to cool (no pipes, no ducts), these are sometimes defined as central systems because they do not work on a room-by-room basis. Moreover their cooling capacity is often much higher exceeding 20TR.

For the purpose of this course, the author defines the central system as those systems which are intended to condition multiple spaces from one base location and are essentially field assembled equipment comprising chillers, air handling units, ductwork, chilled water and condenser water distribution and engineered control system.

PART 2 CENTRAL SYSTEMS v/s LOCAL SYSTEMS

The design of HVAC systems is related to various parameters, including but not limited to:

- Comfort
- Building Architecture
- Building Regulations
- System Controls
- Robustness & Redundancy
- System Flexibility
- System Integration
- Energy Efficiency
- Whole of life costs including capital costs, maintenance costs, energy costs and replacement cost

The above factors represent a minimum set of criteria against which HVAC system selection shall be assessed. User's ultimate objective is to acquire and utilize a HVAC system that will minimize the impact on the natural and physical environment through energy efficiency and simplification of the systems and that will provide the most appropriate approach, on a whole of life basis, in terms of capital, operating, replacement and maintenance costs. In managing conflicting requirements in terms of optimizing the HVAC system selection, the consultant shall prioritize parameters that affect the fit for purpose nature of the system (comfort, reliability). These parameters shall take priority over energy efficiency.

The key facts about centralized and decentralized system are summarized here:

CENTRAL SYSTEMS	DECENTRALIZED SYSTEMS
CONFIGURATION	
A central system is custom-designed for a building and is categorized by field assembly of:	A decentralized system is essentially off shelf, factory assembled, compact equipment consisting of cooling/heat

CENTRAL SYSTEMS	DECENTRALIZED SYSTEMS
• **Source components -** comprising of the <u>compressor</u> (reciprocating, screw, centrifugal or scroll type), <u>condenser</u> (water cooled shell & tube or air cooled finned type), <u>expansion valve</u>, and the <u>evaporator</u> (chilled water shell & tube type or direct expansion finned coil type). All these components are assembled in a skid, known as the chiller package. Refrigerant piping required to connect these parts is also enclosed in this skid. The chiller package is located in a dedicated plant room.	source, distribution, delivery and control functions in a single package.
	The most common Decentralized air-conditioning system includes window, split, package and heat pump air-conditioning units.
• **Distribution system** comprising of chilled water & cooling water pumps, air handling units, and ductwork. The pumps are generally located in the chiller plant room and the air handling units are installed in separate air handling rooms distributed at various locations of the building.	For large buildings decentralized systems may be viewed as collection of multiple independent units placed at different locations in a distributed network with each unit working in isolation. Each system is local self-contained unit consisting of its own compressor/s, evaporator coil, fan, condensing unit and filtration unit.
• **Terminal elements** comprising of grilles, diffusers, ventilation systems, and a number or elements adjusting comfort (local re-heat, humidity treatment, thermostats, air filtering etc.). Heat rejection system (cooling tower/s or air cooled	Decentralized systems maintenance tends to be simple but such maintenance may have to occur directly in occupied spaces.

CENTRAL SYSTEMS	DECENTRALIZED SYSTEMS
condensers) are located outdoors. All these components are field assembled. They perform all the functions as usual similar to a typical refrigeration system; however, all these parts are larger in size and have higher capacities. Central systems allow major equipment components to be isolated in a mechanical room. Another benefit is refrigerant containment. Having the refrigeration equipment installed in a central location minimizes the potential for refrigerant leaks, simplifies refrigerant handling practices, and typically makes it easier to contain a leak if one does occur.	
TYPES	
There are two types of central air-conditioning plants or systems: • Direct Expansion (DX) Type: Here, the air is cooled directly by the refrigerant in the finned type cooling coil of the air handling unit. • Chilled Water (CHW) Type: Here, a secondary cooling medium (chilled water) is used to deliver cooling to one or more locations needing it. The ordinary water or brine solution is chilled to very low temperatures of about 40°F by the refrigeration plant and is pumped to various air	Decentralized systems are essentially direct expansion (DX) type. Depending upon the capacities required and areas served the decentralized equipment category includes: ▪ Window air conditioners; ▪ Residential and light commercial split systems; ▪ Packaged thru-the-wall and window air conditioners; ▪ Self-contained (floor by floor)

CENTRAL SYSTEMS	DECENTRALIZED SYSTEMS
handling units. The chilled water flows through the cooling coil, which cools the air. In comparison to DX systems, the chilled water systems can be easily networked to have multiple air handling units distributed throughout the large distributed buildings while the main chiller package placed at one central location. Chilled water systems are not constrained by distance criteria. Chilled water systems provide greater control flexibility by modulating the chilled water flow rate through the cooling coils served from a single chiller without compromising control on any individual unit.	package systems; ▪ Commercial outdoor roof top packaged systems Since in DX systems, the air is cooled directly by the refrigerant the cooling efficiency is higher. However, it is not always feasible to carry the refrigerant piping to the large distances (beyond 100 ft) therefore the DX type system is usually used for cooling the small buildings or the rooms on the single floor. For this reason, decentralized systems are essentially floor by floor standalone, self contained units each working independent of each other.

HEAT REJECTION OPTIONS

Central air conditioning systems expel heat by air or water cooling. • **Air cooled** - The air cooled method uses finned tube coil condenser. Here the refrigerant flows through the refrigerant piping from evaporator to the condenser. When the refrigerant flows in the refrigeration piping there is lots of drop in its pressure. Due to this the length of the refrigeration tubing and	Most decentralized systems use air-cooled condensers to expel heat. They have to be generally kept very close to the evaporator units and for smaller sized equipment; the length should be 30 to 40 feet whereas for larger systems it may go up to 3 to 4 times this figure. The paucity of good quality soft water makes it imperative to opt for air cooled systems.

CENTRAL SYSTEMS	DECENTRALIZED SYSTEMS
the distance between the condenser and the evaporator should be kept minimum possible. • **Water cooled** - Water cooled systems use shell and tube type condenser. Here, the cooling water is pumped from tubes of the condenser to the cooling tower at high pressure, which is good enough to carry it to relatively long distances. The losses in the pressure of water are accommodated by the sufficient capacity of the pump, which has low capital and running cost. Central system with water cooled heat rejection option thus may virtually be placed at any distance from the cooling equipment. Water cooled units are more efficient and have good overload capacity as these are sized to wet bulb, not dry bulb temperature. At higher ambient dry bulb temperatures, the compressor capacity drops by over 10% for air cooled machines compared to water cooled. In general: ▪ For cooling loads below 100–125 tons, the chiller(s) shall be air-cooled. The capital cost and increased maintenance requirements for a water-cooled	

CENTRAL SYSTEMS	DECENTRALIZED SYSTEMS
system are rarely justified on the cooling loads below 125 tons. • Above 200 tons peak cooling load, the water-cooled systems become justifiable. • Between 100 and 200 tons cooling load, it becomes a matter of the owner's ability to deal with the maintenance requirements of a cooling tower system, and the capital funds available. Note that the central air conditioning systems equipped with centrifugal machines are ONLY available as water-cooled heat rejection option while the reciprocating and screw machines are available with both air-cooled and water cooled options. Poor water quality and regular chemical dosage requirements etc are a limiting factor for water-cooled equipment.	
APPLICATIONS	
The central systems are used when large buildings, hotels, theaters, airports, shopping malls etc are to be air conditioned completely. The largest capacity of chiller available in market is 2000 tons; multiple chillers are installed to cater for higher loads or to create redundancy in operation.	Decentralized systems are more appropriate for low to mid-rise buildings. Also in a building where a large number of spaces may be unoccupied at any given time, such as a dormitory or a motel, decentralized systems may be preferred since these can be totally shut off in the unused

CENTRAL SYSTEMS	DECENTRALIZED SYSTEMS
Often a "hybrid system" which is a combination of a central plant and decentralized packaged units/split units is preferred. For example, a hotel may use packaged unitary air conditioners (or fan coil units served with air-water central system) for the individual guest rooms, roof top units for meeting rooms/restaurants, and a Central plant system for the lobby, corridors and other common spaces.	spaces, thus providing potential energy savings. Decentralized unit capacities range from 0.5 ton to 130 tons (for roof top package units). If decentralized systems are chosen for large buildings, multiple package units may be installed to serve an entire building. This may be an advantage, since each system can be well matched to the interior space that it serves. Decentralized systems can be also be applied for augmenting the cooling needs in the central HVAC systems necessitated due to expansion or addition of more equipment.
USAGE PATTERNS	
Centralized systems are preferred where the usage time is high and consistent.	Decentralized or individual systems are preferred where the air conditioning requirements are low or intermittent. Such systems offer high flexibility in meeting the requirement of different working hours and special design conditions.
ZONING	
Central air conditioning systems may serve	Decentralized (DX) systems are only

CENTRAL SYSTEMS	DECENTRALIZED SYSTEMS
multiple thermal zones* and can have as many points of control as the number of zones. [*A thermal zone is referred to a space or group of spaces within a building with heating and cooling requirements that are sufficiently similar so that desired conditions (e.g. temperature) can be maintained throughout using a single sensor (e.g. thermostat or temperature sensor). Each thermal zone must be 'separately controlled' if conditions conducive to comfort are to be provided by an HVAC system].	suitable for single thermal zone application. The reasoning is as follows: DX systems do not provide modulating control. The capacity control in DX system with fully hermetic sealed compressor is generally accomplished by cycling the compressor ON and OFF in response to the signals from a thermostat. What this means is that the DX system will only have one point of control – typically a thermostat. Thus two rooms with thermostat controllers set at say 22°F and 28°F shall conflict with each other or in other words the two rooms cannot achieve the set conditions unless the rooms are served with independent units. Multiple units are required for multiple zones.

INTERIOR AND EXTERIOR EXPOSURES

Central system can serve both the interior and exterior (perimetric)* zones. Constant volume (CV) type Central systems are suitable for the interior core spaces while the exterior zones are best handled with variable air volume (VAV) type Central system. [* There are two types of zones - perimeter and core zones. The perimeter zones are	Decentralized compact systems are good for exterior (perimetric) areas having large exterior exposure or where spot cooling is required. Interior zones are served by split units which may require extensive plumbing for condensate drain and refrigerant piping.

CENTRAL SYSTEMS	DECENTRALIZED SYSTEMS
highly affected by the external environment and the movement of sun, requiring heating in the winter and cooling in the summer. Perimeter zones extend approximately 15 ft in from the building envelope. Core zones are indoor areas where the heat load is nearly constant and is not influenced by ambient conditions.]	
CAPACITY CONTROL	
Capacity control in central systems (chilled water type) is usually achieved by modulating the chilled water flow rate through the cooling coils served from a single chiller without compromising control on any individual unit. Central chilled water systems are better controlled allowing closer temperature/humidity tolerances in space under almost any load condition.	Decentralized systems do not provide modulating control. The capacity control in decentralized DX system is generally accomplished by cycling the compressor ON and OFF in response to the signals from a thermostat. Typically Decentralized DX systems have a 'fixed' off coil temperature during the cooling mode. (Note - Variable refrigerant flow (VRF) systems are now available, but the success of these systems still need to be time tested.)
ENVIRONMENT CONTROL	
Central systems provide full control over temperature, relative humidity, indoor air quality and air distribution. The quality of air conditioning is much superior and is best suited for applications demanding precise	Decentralized HVAC systems are especially suitable in situations where the absolute highest level of performance is not required. The quality of air conditioning is OK and at justifiable expenditure only thermal air

CENTRAL SYSTEMS	DECENTRALIZED SYSTEMS
control of environmental conditions.	treatment is possible.

TEMPERATURE CONTROL

Central systems allow for proportional control of temperature and eliminate hot spots when the system is properly balanced.	Decentralized system provides simple two position on-off type control. This may lead to high temperature and relative humidity swings.

LOW TEMPERATURE APPLICATIONS

Central chilled water systems have limitations on cold air distribution. The chilled water systems for comfort air conditioning typically operate with a design supply temperature of 40 to 55°F. Antifreeze or brine solutions may be used for chilled water systems (usually process applications) that require supply temperatures below 40°F.	Decentralized systems are better choice for the applications demanding low temperature and low humidity conditions. The application includes the grocery stores, fruit & vegetable stores, meat processing units, instrument rooms, laboratories, bio-medical labs, critical manufacturing and process facilities.

INDOOR AIR QUALITY

Central systems provide excellent dust and particulate air filtration. Multi stage filtration can be employed to improve the quality of supply air and the fan static pressure can be selected to suit the pressure drop. These systems can incorporate high efficiency particulate filters (HEPA), which offer 99.99% efficiency down to 0.3 micron.	Decentralized systems cannot be modified to include high level of filtration due to fan static pressure limitations. Decentralized systems are typically available at 2 inch-wg pressure (max.) which is often inadequate to overcome the filter resistance.

FRESH AIR VENTILATION CONTROL

CENTRAL SYSTEMS	DECENTRALIZED SYSTEMS
Central systems provide good control over ventilation air and allow for fixed or varying quantities of fresh air.	Decentralized systems do not provide much flexibility on the control of fresh air.
INDIVIDUALIZED CONTROL	
In a central system, the individual control option is not always available. If individual control is desired, the system shall be designed as variable air volume system (VAV). A variable air volume (VAV) system primarily alters the air delivery rates while keeping the fixed off-coil temperatures. Constant air volume (CAV) systems alter the temperature while keeping the constant air delivery. CAV systems serving multiple zones rely on reheat coils to control the delivered cooling. This incurs lot of energy wastage due to simultaneous cooling and heating.	Decentralized systems offer room-by-room control, which provides greater occupant comfort through totally individualized control options -- if one room needs heating while an adjacent one needs cooling, two decentralized systems can respond without conflict. Heating and cooling capability can be provided at all times, independent of other spaces in the building.
EFFICIENCIES	
Central HVAC systems deliver improved efficiency and lower first cost by sharing load capacity across an entire building. A central chilled water system using high efficiency water cooled chillers typically provide greater energy efficiency, but efficiency and stability of operation is compromised when only a small proportion	Decentralized systems have high kW per ton compared to chiller systems. But in buildings where a large number of spaces may be unoccupied at any given time, the units may be totally shut off in the unused spaces thus providing potential energy savings.

CENTRAL SYSTEMS	DECENTRALIZED SYSTEMS
of space is using air conditioning.	Federal law mandates a minimum efficiency of 10 SEER* for both split and packaged equipment of less than 65,000 Btu/h capacities. The American Society of Heating, Refrigeration and Air Conditioning Engineers (ASHRAE) recommend 10 EER* for equipment between 65,000 and 135,000 Btuh. ASHRAE standard 90.1 recommends other efficiencies for larger equipment. It is often cost effective to pay for more efficient equipment. For example, upgrading from a 10 SEER to a 12 will reduce cooling costs by about 15 percent. Upgrading from a 10 to a 15 reduces cooling costs by about 30 percent.

Chiller efficiency is typically defined in terms of kW/ton and/or its coefficient of performance (COP). The COP is the ratio of output BTU's divided by the input BTU's. If the nominal rating of the chiller is 1 ton of refrigeration capacity, equivalent to 12,000 Btu/hr output, and the input energy is equivalent to 1 kW, or 3,413 Btu/hr, the resulting COP is 12,000/3,413 or 3.52.

- **Reciprocating** chillers have a peak load power requirement of 1.0–1.3kW/ton, depending on capacity and ambient air temperature.

- **Screw** chillers have a peak load power requirement of 0.5–0.7kW/ton.

- **Centrifugal** chillers are most efficient at peak load and they consume the least power (kW per ton) at full load operation. At ARI standard rating conditions centrifugal chiller's performance at full design capacity ranges from 0.53 kW per ton for capacities exceeding 300 tons and between 0.6 to 0.7 kW per ton for capacities up to 300 tons.

[In a building where a large number of spaces may be unoccupied at any given time, the central system operating at part

Efficiency Terms

- SEER – The Seasonal Energy Efficiency Ratio is a representation of the cooling season efficiency of a heat pump or air conditioner in cooler climates. It applies to units of less than 65,000 Btu/h capacities. The higher the SEER rating, the more efficient the AC system operates.

- EER – The Energy Efficiency Ratio is a measure of a unit's efficiency at full load conditions

CENTRAL SYSTEMS	DECENTRALIZED SYSTEMS
load will consume higher energy (kW/ton). During design phase it is necessary to select optimum configuration of chiller machines; for instance a peak load of 225 tons could be served through 3 x 75 ton machine so that one machine can be switched off at low loads. Alternatively, central chilled water system can incorporate decentralized systems for areas requiring 24hrs operation such as server rooms, data centers etc.]	and 95 degrees outdoor temperatures. It typically applies to larger units over 65,000 Btu/h capacities. • Btu/h – Btu/h is a rate of heating or cooling expressed in terms of British Thermal Units per Hour. • Ton – One ton of cooling is the energy required to melt one ton of ice in one hour. One ton = 12,000 Btu/h.

AIR DISTRIBUTION SYSTEM

Central systems are characterized by: • High pressure loss in the distribution system; • High area requirements for air distribution system; • High efficiency of fans.	Decentralized systems are characterized by small static pressure of fans and low efficiency of fans. Decentralized systems air distribution is not as good as ducted systems.

CONDENSATE REMOVAL

Condensate removal is easily achieved in central systems since the cooling coil (evaporator) is located remotely in air handling unit room.	Condensate disposal is cumbersome and sometimes difficult especially in multiple unit installation.

SYSTEM FLEXIBILITY

Chilled water systems are the engineered systems that are generally supplied as the	Decentralized systems usually provide fixed air delivery rate of 400 cubic feet

CENTRAL SYSTEMS	DECENTRALIZED SYSTEMS
custom built units. These can be fabricated to suit the designer application and the air delivery rate can be sized irrespective of the refrigeration capacity. The cooling coils in a central plant can be custom designed to handle higher latent loads and thus provide better control over moisture. The cooling coils can be selected for high rows deep (6 or 8 row deep coil provide enhanced surface area) for effective condensate removal.	per minute (cfm) per ton of refrigeration. Decentralized systems cannot be networked conveniently. The refrigerant piping plays a key role in connection of various components in terms of size, length and pressure drop. Split units installation is restricted by distance criteria between the condensing unit and the evaporator, which is usually 30 to 40 feet for smaller units and around 100 to 120 feet for larger units. The size of the cooling coil and condenser coils is standard generally factory fixed and is typically 3 or 4 row deep.
POSITIVE PRESSURIZATION	
It is possible to maintain positive pressure with central systems. The supply air quantities of central system can be designed to any value by incorporating custom build fans.	Small compact decentralized systems are generally 100% re-circulation type and may not be suitable for the applications requiring high air delivery rates and the areas requiring significant positive pressurization. A standard unitary system provides 400 CFM of air delivery capacity per ton of refrigeration.
CROSS CONTAMINATION	

CENTRAL SYSTEMS	DECENTRALIZED SYSTEMS
Central systems require considerable engineering effort to keep supply and return system independent for areas requiring separation. Independent air handling units (AHU) may be required for critical areas where cross contamination is a concern.	It is easy to provide independent package units where cross-contamination is a concern. Application includes food courts, laboratories, hotels restaurants, hospitals etc.
AESTHETICS	
Central systems are generally designed as concealed systems and can be easily blended with the aesthetics.	The decentralized units can be unappealing and may not necessarily blend well with the aesthetics. Window or wall through package units for example, must penetrate vertical elements of the building envelope -- with substantial impact on appearance and envelope integrity.
LOAD SHARING	
Central systems permit building-wide load sharing; this may result in reduced equipment sizes (and costs) and the ability to shift conditioning energy from one part of a building to another.	Lack of interconnection between multiple compact units means that loads cannot be shared on a building-wide basis. A peak load capacity shall be provided for each zone. The capacity of Decentralized unit equipment needs to be determined for peak load of the zone without any diversity.
OPERATIONAL RESOURCES	
Trained and skilled operators are required to operate central systems due to	Decentralized systems are easy to operate and are essentially plug and

CENTRAL SYSTEMS	DECENTRALIZED SYSTEMS
complexity of controls and numerous field assembled items interfacing with each other.	play type. Operation and maintenance of decentralized units tends to be simple and available through numerous service providers.
LIFE EXPECTANCY	
Central systems have longer life. The economic life for reciprocating compressor chillers is normally 15 years, while screw and centrifugal chillers have an expected economic life of 25 years.	Decentralized systems generally have a much shorter useful life (8-10 years).
ECONOMY OF SCALE	
Central air conditioning systems offer opportunities of economies of scale. Larger capacity refrigeration equipment is usually more efficient than smaller capacity equipment and require lower capital expenditure over 100TR.	Decentralized systems do not benefit from economies of scale. Capital costs and the operating costs generally tend to be higher for larger setups requiring 100TR or more.
MAINTENANCE	
Grouping and isolating key operating components in mechanical room allows maintenance to occur with limited disruption to building functions.	Decentralized systems maintenance tends to be simple but such maintenance may have to occur directly in occupied spaces.
SERVICING	
Central air conditioning systems are highly sophisticated applications of the air	Decentralized systems are not complicated by interconnections with

CENTRAL SYSTEMS	DECENTRALIZED SYSTEMS
conditioning systems and many a times they tend to be complicated. As system size and sophistication increase, servicing and replacement may become more difficult and may be available from specialist providers.	other units; service generally readily available and comfort can be quickly restored by replacing defective chassis and is available through numerous service providers.
MECHANICAL ROOM SPACE	
A central plant will require considerable floor space to house chilling machines, chilled water and condenser water pumps, electric and control panels. The chiller plant room will require an adequate height of 4.3 to 4.9 meters for installation and servicing. In addition mechanical space is required for housing air handling units at various locations in large building. This typically can range from 4 to 6% of the overall building foot print.	Decentralized systems do not require any form of plant room/area within the building. Refrigeration package is integral to the package unit/condensing unit which is generally located outdoors.
PLENUM SPACE	
Central systems need plenum space above false ceiling to accommodate the air distribution system comprising of ductwork and auxiliaries (dampers, attenuators and fittings). This results in an increase in floor-to-floor height and consequent building cost. Additional expenses are also necessary for provision of false ceiling to hide the air distribution system.	Decentralized systems can be arranged without false ceiling or plenum space. By saving the false ceiling void the resulting building slab to slab height can be lowered by almost 20%.

CENTRAL SYSTEMS	DECENTRALIZED SYSTEMS
CORE/SHAFT SPACE	
A shaft is needed to house chilled water piping, condenser water piping, supply, return and fresh air ducts and power/control distribution cables.	Decentralized systems do not require chilled water pipes. Small bore refrigerant piping can easily be taken through wall/floor and attic space.
ENERGY MONITORING	
Central systems do not provide flexibility of individual energy metering very easily. Air conditioning costs can only be apportioned on an overall basis.	The energy utilization of decentralized compact units can be simply measured by installing a decentralized energy meter with each unit.
A complex metering system generally based on BTU/hr (measured from flow and temperature differential) of chilled water energy is first measured to convert to equivalent power units in kWH.	If the tenants are paying the utility bills, multiple compact units may make it easier to track energy use, as only the specific unit serving that tenant would be used to meet the individual cooling requirements.
STRUCTURAL DESIGN/COSTS	
For Central systems, the building structure should be designed to take the weight of equipment.	The decentralized systems are smaller in size and are less bulky.
Suitable vibration control must be considered.	Costs are lower due to less assembly of component ducting etc. However interference to the façade is high.
Adequate load bearing beams and columns must be available for lifting and shifting of such equipment.	
CONDENSATE DRAIN	

CENTRAL SYSTEMS	DECENTRALIZED SYSTEMS
Central systems require plumbing and drainage arrangement in the plant room where cooling water pumps are located and also in mechanical rooms where AHU/FCU cooling coils are provided.	Since majority of time the evaporator unit is located with in the zone or at the zone boundary, the plumbing need to be carried out in the indoor spaces.
NOISE	
Since mechanical room is located away from the conditioned space, the machine noise is reduced. If the air handling mechanical room is located indoors, the room walls must be acoustically treated.	Decentralized equipment is generally located inside, adjacent or closer to conditioned space. Operating sound levels are noticeable although tolerable.
LOAD DIVERSITY	
Central systems can be designed zone wise with significant diversity (70-80%) in overall plant load capacity.	The decentralized systems being small are designed for full peak load. No diversity is taken on design.
MECHANICAL ROOM ACCESS	
Thought must be given to the access path to plant rooms and AHU rooms. The equipment may be quite bulky and voluminous. In case of a breakdown, the machine may have to be shifted to a service shop for repair. The building design must provide this space.	The decentralized systems are usually compact. Replacement is quite simple and easy.
SMOKE CONTROL	
It is possible to design central system to	Decentralized systems are standard

CENTRAL SYSTEMS	DECENTRALIZED SYSTEMS
include active smoke control and building pressurization. This is best accomplished by "all-air" HVAC system.	items and may not suit modifications other than interlocking the fan motors with fire detectors/panel.

BUILDING MANAGEMENT SYSTEMS

Central systems are amenable to centralized energy management systems. If properly managed these can help in optimal utilization of the air conditioning plant and can reduce building energy consumption besides providing effective indoor temperature and humidity control.	Decentralized system units cannot be easily connected together to permit centralized energy management operations. Decentralized systems can be integrated to BMS with respect to on-off functions.

POSSIBILITIES OF HEAT RECOVERY

With central systems it is possible to incorporate strategies which are desirable with increased ventilation rates: ✓ Increased re-circulation with high efficiency filters ✓ Heat recovery devices can be provided ✓ Economizer: An economizer allows outside air to be used for cooling when its temperature is lower than the temperature inside the building. The economizer function is part of the control package on many single-packaged units. ✓ Automatic carbon dioxide monitoring can be incorporated.	Decentralized systems are not amenable to heat recovery devices (such as night-setback or economizer operation) is not possible.

CENTRAL SYSTEMS	DECENTRALIZED SYSTEMS
THERMAL ENERGY STORAGE	
Central systems can be applied with large thermal energy storage systems to take benefits of reduced cooling demand during expensive peak load periods.	With decentralized systems, switching off few of the multiple units can control the peak load energy demand. Thermal energy storage is not economical with compact systems.
RELIABILITY	
Central systems are categorized as non-distributed systems. As a non-distributed system, failure of any key equipment component (such as pump or chiller) may affect an entire building. Standby equipment needs to be perceived during design.	Decentralized systems tend to be distributed that increases reliability; a building conditioned using DX system may have a dozen or hundred of individual and independent units located throughout the building. Failure of one or two of the units may not impact the entire building. On a smaller scale this may be viewed as a disadvantage unless standby is provided.
REDUNDANCY	
Central systems provide greater redundancy and flexibility as it is easy to provide a standby chiller and pump in the same plant room. Either of the chillers (main & standby) can act as standby to any of the air-handling units (main & standby). A multiple chiller plant arranged in N+1 configuration provides more than 100% redundancy because of the fact that most of	In the decentralized DX system one compressor is associated with one air-handling unit cooling coil, hence the flexibility & redundancy of operation is limited. It is not always possible to provide a non-working standby unit. Therefore whenever a unit suffers a breakdown, air conditioning is inadequate causing

CENTRAL SYSTEMS	DECENTRALIZED SYSTEMS
the chiller plant operates at off-design conditions 99% of the time.	user complaints. Decentralized rooftop units or package units are often provided with standby.
PROCUREMENT	
Central systems are procured from multiple vendors for instance chiller, boilers, pumps, cooling tower, expansion vessel, air handling units, acoustic silencers, piping, ducting & auxiliaries etc. System designer has to produce schematic, layout, control philosophy and general arrangement drawings to assist installation. Delivery of source and distribution equipment is longer.	One manufacturer is responsible for the final unit. Manufacturer-matched components have certified ratings and performance data. Factory assembly allows improved quality control and reliability. The decentralized units are easy to install, which means less mess, or disruption or downtime. Offer short delivery schedules and generally available as off the shelf item.
CONSTRUCTION TIME	
Design, engineering and construction of central systems take longer time.	Decentralized systems are compact and offer much simpler, faster, and less expensive installation.
OWNERSHIP	
Most landmark buildings with a single corporate or government owner have a preference for a central plant, as the quality of air conditioning is far superior and life expectancy is higher. The operation and maintenance costs are also lower than de-Centralized floor-by-floor system.	A multiple owner facility requires a system, which provides individual ownership and energy billing for which a decentralized floor-by-floor air conditioning system is most suited subject to economics of space and aesthetics.

CENTRAL SYSTEMS	DECENTRALIZED SYSTEMS
	Decentralized systems provide greater flexibility of remodeling the space as areas are leased and occupied.

CAPITAL COSTS

CENTRAL SYSTEMS	DECENTRALIZED SYSTEMS
The initial cost of a central air conditioning system is much higher than a decentralized system. Depending on the type of equipment, the cost can vary to a great extent. For example, a reciprocating packaged chiller is much cheaper than a screw-packaged chiller and the screw-packaged chiller is cheaper than a centrifugal chiller. The capital cost expressed in dollars per ton is generally lowest for reciprocating and highest for screw compressors. Centrifugal chillers cost lower than screw chillers by 10 to 15% in most sizes at the same operating conditions. First cost of centrifugal chiller is higher than reciprocating under 200 tons but becomes competitive in the larger sizes. Air-cooled machines are costlier than water-cooled machines. Therefore, the budget available with the owner at the time of building the facility play a major role in selecting the type air conditioning system. VAVs and a building management system if added will increase the capital cost by 10%-15%. However there will be a saving in power cost and so it is important to work	Decentralized systems almost always have a much lower acquisition cost than the total cost to design and purchase the components for an equivalent custom designed central system. Lower installation costs provide additional savings. One other most common reason for selecting a decentralized system is low installed cost. It requires less field labor and has fewer materials to install.

CENTRAL SYSTEMS	DECENTRALIZED SYSTEMS
out the payback period to determine the techno-commercial liabilities of the final selected system.	

ENGINEERING COSTS

Central chilled water systems incur around 4 to 5% of the capital costs towards engineering efforts. • A central plant equipment, ducting, piping layouts and control schemes are much more complex. • Layout finalization is time consuming and requires close interaction with architect, interior layouts, electrical and structural disciplines. • The system selection must precede the final architectural design of the building. Even though such engineering inputs seem to add to the cost and time of the project.	Engineering costs, skills, time and risk factors for designing and installing decentralized floor-by-floor system are usually lower than those for a central system for the following reasons: • Load calculations and corresponding equipment selections are less critical. The multiple numbers of modular units will provide built in safety cum flexibility into the design. • Since the units are factory built standard equipment, the quantum of work to be carried out at site is much less as compared to central system. • Layouts are much simpler and repeated multiple times.

INSTALLATION COSTS

The installation cost of a central plant is much higher because • Main air conditioning equipment is heavy and voluminous requiring	Decentralized system provides simple and faster installation. These are easy to install and less time consuming since standard size units are readily

CENTRAL SYSTEMS	DECENTRALIZED SYSTEMS
strong foundations, heavy lifting and proper material handling facility at site. • Some equipment requires extra care during installation to ensure minimum vibrations and smooth operation. • Larger quantities of ducting, piping, insulation and false ceiling are required.	available. Replacements can be carried out very fast.

OPERATING COSTS

The modern centrifugal machine is capable of operating at a power consumption of 0.50 - 0.60 kW per ton. In addition to the above, centrifugal machines are now available with variable speed drives (VSD), which enables machines to operate at off design conditions at 0.40, 0.30 and even at 0.20 kW/ton. This leads to an unprecedented energy saving. Note: For all air-conditioning systems a vast majority of operating hours are spent at off design conditions. Therefore it is important select machines which the best off design performance.	The power consumption of Decentralized compact units can vary from 1.0 kW per ton to 1.3 kW per ton. The type of compressors used in these machines is either hermetic reciprocating type or scroll. The part load efficiency of such units is lower than their full load efficiency. Cooling efficiency for air conditioners, splits, package units and heat pumps is indicated by a SEER (Seasonal Energy Efficiency Ratio) rating. In general, the higher the SEER rating, the less electricity the unit will use to cool the space. The government-mandated minimum efficiency standards for units installed in new homes at 10.0 SEER. Air conditioners

CENTRAL SYSTEMS	DECENTRALIZED SYSTEMS
	and heat pumps manufactured today have SEER ratings that range from 10.0 to about 17.
MAINTENANCE COSTS	
The breakdown, repair, replacement and maintenance cost of central plants can be expensive and time consuming. However, the frequency of such breakdown is quite low. These systems require routine inspection and planned checks. Daily operation also adds to the running cost, as trained operators are required. Maintenance costs are difficult to predict since they can vary widely depending on the type of system, the owner's perception of what is needed, the proximity of skilled labor and the labor rates in the area. A recent survey of office buildings indicated a median cost of $0.24 per sq. ft per year.	The decentralized system repair cost per breakdown is normally low. With the emergence of reliable hermetic and scroll compressors, their maintenance expenditure has shown remarkable improvements and is less time consuming and simple. Roof mounted packaged units typically have maintenance costs of 11% or higher than a central plant system serving the same building.

In nutshell central systems provides better comfort conditions, quality of indoor parameters and energy efficiency. The decentralized systems are suitable for small or medium sized buildings free of multiple thermal zones and demanding 100 TR or less of air-conditioning.

From energy efficiency point of view it is highly recommended to evaluate your selection thoroughly if any of the conditions below are true.

1. If the building floor area exceeds 10000 sq-ft;

2. Ratio of occupancy hours to operative hours of less that 0.6;

3. Annual energy consumption in excess of 50,000 BTU/sq-ft;

4. Total capacity of heating and cooling equipment combined capacity exceeding 100 BTUH/sq-ft.

Course Summary

In commercial workplaces the comfort, safety and productivity of the occupants is affected by poor performance of HVAC systems, which has indirect cost implications. There are several choices for the type of air conditioning systems, each satisfying the HVAC objectives with different degrees of success. In general central systems provide better quality of indoor parameters and energy efficiency. However, central systems are costly to build but the operating costs tend to be low on large systems. The decentralized systems are suitable for small or medium sized buildings free of multiple thermal zones and demanding 100 TR or less of air-conditioning. For intermittent use buildings there is a growing trend to select a combination of central plant and packaged or split units to meet the overall functional requirement of the buildings.

With the strong trend in the Heating, ventilation and Air-conditioning (HVAC) industry emphasizing energy savings, there is an equally a concern from the owners & operators that the installed costs of new and replacement systems be as economical as possible. In applying this concept to the buildings, the designer should consider not only the first costs but also the maintenance costs, rehabilitation costs, user costs, and reconstruction costs. The final choice of an HVAC system, whether central or floor-by-floor is a critical decision required to be taken before the facility design is completed. The team consisting of architect and HVAC design engineer need to integrate the user's requirements and the building functional requirements. The finally selected system must fit in to the owner's capital budget and anticipated life cycle operation and maintenance cost.